634.93078 JOS
Joslin, Les, author.
US Forest Service ranger stations of the West

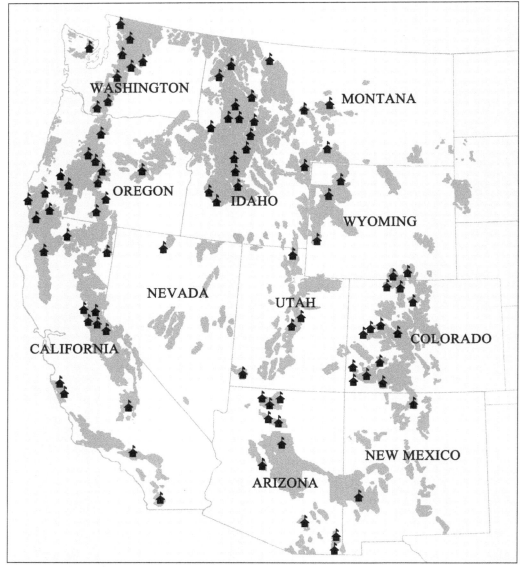

Ninety of the 92 historic US Forest Service ranger stations in the West pictured in this book are shown on this map of the 11 western states. A ranger station in Petersburg, Alaska, and the ranger boat *Chugach* moored there are the other two. Petersburg is shown on the Alaska Region map on page 123. (Map by Gary Asher and Les Joslin.)

ON THE COVER: A Forest Service ranger stands in front of the new Silver Creek Ranger Station log office building in the Snoqualmie National Forest in Washington state in 1931. (Photograph by Fred W. Cleator, courtesy of the US Forest Service and Rick McClure.)

IMAGES *of America*

US FOREST SERVICE RANGER STATIONS OF THE WEST

Les Joslin

ARCADIA
PUBLISHING

Copyright © 2019 by Les Joslin
ISBN 978-1-4671-0315-2

Published by Arcadia Publishing
Charleston, South Carolina

Library of Congress Control Number: 2018964821

For all general information, please contact Arcadia Publishing:
Telephone 843-853-2070
Fax 843-853-0044
E-mail sales@arcadiapublishing.com
For customer service and orders:
Toll-Free 1-888-313-2665

Visit us on the Internet at www.arcadiapublishing.com

Contents

Acknowledgments		6
Introduction		7
1.	Northern Region	9
2.	Rocky Mountain Region	29
3.	Southwestern Region	49
4.	Intermountain Region	65
5.	Pacific Southwest Region	81
6.	Pacific Northwest Region	97
7.	Alaska Region	123
Epilogue		126
Bibliography		127

ACKNOWLEDGMENTS

This effort to share with a wider readership some of the results of three decades of research and writing about historic Forest Service ranger stations in the western United States, through seven regional chapters and almost 200 captioned photographs, reflects the contributions of literally hundreds of sources of information and images that cannot be acknowledged individually in this small space.

Foremost among these are the many pioneer Forest Service rangers who wrote about and photographed—often anonymously—their storied agency's early and usually remote stations. Close on their heels come today's Forest Service archaeologists, historians, and others who appreciate their agency's heritage and have kindly shared the results of their work to benefit my research and writing projects.

Ultimate thanks are due to the late-19th- and early-20th-century visionaries of the American conservation movement, whose many successes included the establishment of the Forest Service and the National Forest System. No lesser body than the Congress of the United States deserves credit for passing the many laws that not only advanced accomplishment of the Forest Service mission, but also funded construction and preservation of the ranger stations from which that mission was pursued. Among these are the Civilian Conservation Corps (CCC) Reforestation Relief Act of 1933 and the Emergency Relief Appropriation Act of 1935, which established the CCC—Pres. Franklin D. Roosevelt's "Tree Army," which, from 1933 until 1942, provided employment to three million and accomplished much good work—and funded Great Depression construction of Forest Service ranger stations and other facilities, as well as the Federal Lands Recreation Enhancement Act of 2004 and the American Recovery and Reinvestment Act of 2009, which helped fund restoration of historic ranger stations as heritage and recreation resources.

More immediately, thanks again to fellow writer Tor Hanson of Bend, Oregon, who reprised the same technical assistance reflected in the compilation and integration of the images that illustrate this book, which he provided to my two previous Arcadia Publishing titles, and to Jody Conners of Maverick Publications of Bend, Oregon, for assistance with image recovery. Thanks also go to Arcadia Publishing for embracing this project.

And, of course, thanks to my wife, Pat, for her understanding of and patience with yet another writing project.

INTRODUCTION

Among US government officers and offices, forest rangers and ranger stations are more romantic than most. Both are symbols of the American West's last frontier shared with the cattlemen, sheepmen, lumbermen, miners, homesteaders, and others who used—and in many cases, still use—the public forests and rangelands to build and sustain economies and ways of life.

The first forest rangers, appointed in 1898, worked for the General Land Office, the US Department of the Interior agency charged with looking after public lands including the forest reserves set aside in the West beginning in 1891. After the establishment of the Forest Service in the US Department of Agriculture and transfer of forest reserve administration to that new agency in 1905, Uncle Sam's forest rangers became Forest Service rangers. In 1907, the forest reserves became national forests.

Gifford Pinchot, the founding chief of the Forest Service—the title was "forester" at the time—accepted the better Interior Department rangers into his new Forest Service. Political appointment of rangers gave way, in Pinchot's new outfit, to hiring under US Civil Service Commission rules. This helped Pinchot recruit the caliber of men needed to earn the trust and respect of the American people. As Pinchot wrote in *Breaking New Ground*, his 1947 autobiography, the ranger "was the officer with whom the [public] did most of their Forest Reserve business. On how he handled himself and them depended mainly their attitude toward the Reserve and the Service."

The first Forest Service manual, a pocket-sized volume called *The Use of the National Forest Reserves*, or *Use Book* for short, specified that every applicant for a ranger job "must be, first of all, thoroughly sound and able-bodied, capable of enduring hardships and of performing severe labor under trying conditions." In the next sentence, which became famous, the *Use Book* pointed out that "Invalids seeking light out-of-door work need not apply." A ranger "must be able to take care of himself and his horses in regions remote from settlements and supplies. He must be able to build trails and cabins, ride, pack, and deal tactfully with all classes of people. He must know something about land surveying, estimating and scaling timber, logging, land laws, mining, and livestock business." That was a tall order for a man also required to "own and maintain his own saddle and pack horses," all for $900 a year.

That order was filled by an "examination of applicants along [those] practical lines" taken by men whose motives were as diverse as their backgrounds. After a written test eliminated the illiterates, a practical skills test made sure the Forest Service hired men who could do the job in the woods. From the beginning, the small and scattered force of early-day forest rangers patrolled and protected vast reaches of mountainous forests and rangelands under often adverse conditions. In so doing, they won over the American public to the national forest idea. By the time Pinchot finished as forester in 1910, most opposition to Forest Service management of the national forests was past and a second generation of rangers—many with college degrees—was wearing the pine tree shield.

As rough and ready as these early forest rangers—and their wives—were, they still needed places in which to live and from which to work. They seem to have recognized that before their supervisors did, and some early ranger stations were built. In the 1907 edition of the *Use Book*, Pinchot's office promised ranger stations that would promote a specific image of the Forest Service. "Whenever possible," the book specified—in perhaps the first stab at standardizing ranger station architecture—those "cabins should be built of logs, with shingle or shake roofs." And the *Use Book* continued, "Cabins should be of sufficient size to afford comfortable living . . . and the ranger shall be held responsible for the proper care of the cabin and the ground surrounding it. It is impossible to insist on proper care of camps if the forest officers themselves do not keep their cabins as models of neatness." A tradition, as well as an image, was being forged.

That tradition and that image—as shown in these pages—evolved over the ensuing century as Forest Service needs for ranger stations and other facilities throughout the National Forest System evolved and opportunities to construct them—especially Great Depression era labor and funding—occurred.

In keeping with this guidance and small budgets, the earliest Forest Service ranger stations normally consisted of a single cabin and, perhaps, a barn along with another outbuilding or two and a corral. Pinchot's guidance was clear enough, but various interpretations of new regulations by field personnel working in diverse environments thousands of miles from the Washington office often resulted in a lack of uniformity in field operations—including planning and construction of ranger stations. From the beginning, ranger stations varied from one part of the West to another. In the arid Southwest, for example, ranger station builders often used such local materials as adobe brick and stone. And, of course, some ranger stations were previously existing buildings—some of them abandoned miners' cabins and cowboys' line shacks—pressed into service. Nevertheless, as specified in the *Use Book*, early ranger station architecture was epitomized by the simple log cabin. As time passed, other factors affected how and where the Forest Service built ranger stations. These included the administrative decentralization of the agency, changes in national forest management, advances in transportation and communications, and access to construction resources.

In 1908, three years after it was established, the Forest Service decentralized. Six district offices under district foresters—renamed regional offices under regional foresters in 1930—at Missoula, Denver, Albuquerque, Ogden, San Francisco, and Portland were set up between the Washington office and the western national forests. Also in 1908, the national forests were subdivided into ranger districts, still the basic administrative unit of the National Forest System. District rangers were soon assigned, and ranger stations were their headquarters. District rangers reported to forest supervisors, who in turn reported to their district—later regional—foresters, who answered to the forester in Washington, DC.

Most ranger district staffs remained small during the Forest Service's first few decades. A ranger, who did most of his district's work himself, usually was aided by a clerk and perhaps a few seasonal assistants, such as fire guards. Ranger stations were small, and even smaller satellite facilities called guard stations housed forest guards and provided shelter for the ranger when in the field. Many rangers of this era did not occupy their stations year-round but were detailed to their supervisor's office or even to the district—later regional—forester's office during the winter. But their small ranger stations, where they worked with the public during the summer field season if not all year, became Forest Service symbols.

Today, a diverse collection of historic Forest Service ranger stations constructed during the agency's first 50 years—many of them during the Great Depression by CCC labor with emergency funding—remains in the national forests of the West to reflect the agency's history. Ninety-two of them are pictured and interpreted in this book's chapters, defined by Forest Service administrative regions.

One
NORTHERN REGION

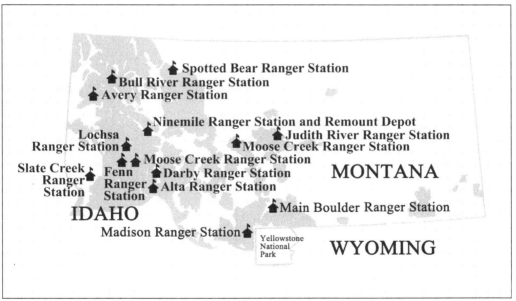

The national forests of the Northern Region comprise 25 million diverse acres that stretch from the rugged peaks and timbered canyons of northeastern Washington, northern Idaho, and western Montana, through eastern Montana's rolling hills and isolated ponderosa pine woodlands, and into the prairies and badlands of the Dakotas. Region 1 is, in many ways, the "culture hearth" of the Forest Service. Many of its influential leaders made their marks there. Many of its pivotal events occurred there. And America's first ranger station, Alta Ranger Station in the old Bitter Root Forest Reserve, was built there in 1899. (Map by Gary Asher and Les Joslin.)

US Department of the Interior General Land Office forest rangers Nathan W. "Than" Wilkerson (left) and Henry C. "Hank" Tuttle raise the American flag at their new Alta Ranger Station on July 4, 1899, before the forest reserves were transferred to the US Department of Agriculture's new Forest Service in 1905 and renamed national forests in 1907. (Photograph by N.W. Wilkerson, courtesy of the US Forest Service.)

Rangers Wilkerson and Tuttle needed shelter in the wilderness where they did Uncle Sam's work of preventing timber theft and fighting forest fires and, out of necessity, built the rustic, sod-roofed, one-room Alta Ranger Station cabin on their own initiative and at their own expense. For over a century, agency and civic pride have ensured its preservation. (Photograph by the author.)

Historic Alta Ranger Station, about 30 miles from Darby, Montana, is identified as "the first Forest Service ranger station" by this Forest Service sign. "In 1904 this site reverted to private ownership under a mining claim," the sign reads. "The Lion's Club of Hamilton, Montana, purchased the site in 1941 and donated it to the Forest Service for preservation as a Montana Historic Landmark." (Photograph by the author.)

In 1905, the year the Forest Service was founded, ranger Harry S. Kaufman built the Main Boulder Ranger Station north of Yellowstone National Park in the Absaroka division of the Yellowstone Forest Reserve, which in 1907 became the Absaroka National Forest in Montana. In 1945, the Absaroka was consolidated into the Gallatin National Forest. These rangers saddled up there about 1920. (Courtesy of the US Forest Service.)

Telephone communications were important to early forest rangers' work. Ranger Harry Kaufman places a call from a field telephone around 1930 (left) and communicated with the Main Boulder Ranger Station (right). (Both, courtesy of the US Forest Service.)

A young helper known as the "Coyote Kid" and ranger Granville "Granny" Gordon's family—his wife, Pauline, and their daughters—stand in front of the Bull River Ranger Station in the old Cabinet National Forest in northwestern Montana in 1908. Except for the flag, this ranger station looked like a typical Montana mountain homestead complete with an orchard, a garden, and outbuildings. (Courtesy of the US Forest Service.)

Historic Bull River Ranger Station, completed in 1908, in service as a ranger district headquarters until 1920 and then as a guard station until the 1970s, has been maintained and restored over the years by citizen volunteers and Forest Service personnel. It became a Kootenai National Forest recreation rental that sleeps up to eight visitors. (Photograph by the author.)

Ranger Thomas Guy Meyers parked his Forest Service "rig" in front of the Judith River Ranger Station about 1915. He built this Lewis and Clark National Forest district headquarters in 1908, which served as his home and office in that central Montana ranger district until 1931 and the station's log horse barn in 1909. (Courtesy of Marian Jeanne Setter and the US Forest Service.)

Meyers, his wife, Emily, and their son Robert relax on the front porch of the Judith River Ranger Station about 1915. After 1931, the structure was used seasonally as a guard station by Forest Service fire, timber, and trail crews until 1981. (Courtesy of Marian Jeanne Setter and the US Forest Service.)

The old Judith River Ranger Station deteriorated until, in April 1992, the new Northern Region Historic Preservation Team and Lewis and Clark National Forest personnel and volunteers began the structure's restoration by replacing its shingle roof. (Courtesy of the US Forest Service.)

Continued maintenance and additional restoration work prior to the Forest Service Centennial in 2005 kept the historic Judith River Ranger Station in good condition as a heritage site and available to the public as a year-round recreation rental. (Photograph by the author.)

Moose Creek Ranger Station, built in the Helena National Forest in Montana in 1908, was a district ranger's headquarters for about 20 years. After supporting CCC work during the Great Depression and US Army war dog training during World War II at Camp Rimini, the cabin was sold to private owners. About 50 years later, it returned to Forest Service ownership and, by 2005, was restored as a summer education facility and a winter rental cabin. (Courtesy of the US Forest Service.)

The old St. Joe National Forest's historic Avery Ranger Station in Avery, Idaho, comprises this 1909 log office building, two early 1920s log residences, and a 1928 bunkhouse. The new ranger station office building almost fell victim to the 1910 forest fires that devastated much of northern Idaho and western Montana. After the fires, it became the center of a station that included a large bunkhouse, a cookhouse that fed many hungry mouths before it closed in 1960 and was torn down in 1970, and a large barn that supported over a hundred pack mules. (Photograph by the author.)

Frank Hartman, second ranger of the Slate Creek Ranger District, Nez Perce National Forest, from 1911 to 1915, lived with his family in this two-story log cabin five miles up Slate Creek from its current location at today's Slate Creek Ranger Station on US Highway 95 about 20 miles south of Grangeville, Idaho. (Courtesy of the US Forest Service.)

The historic Slate Creek Ranger Station cabin was dismantled by the Nez Perce Hotshots and moved to its current location in 1975. After it was reassembled and given a new floor and shake roof, it was set up to interpret the life of an early Forest Service ranger and his family to forest visitors. (Photograph by the author.)

Historic Moose Creek Ranger Station was until 1995 the summer headquarters of the National Forest System's only entirely wilderness ranger district. At about 560,000 acres, the Moose Creek Ranger District then included the Nez Perce National Forest portion of Idaho's 1.3 million–acre Selway-Bitterroot Wilderness sprawled across seven ranger districts in four national forests. (Courtesy of the US Forest Service.)

Newly married district ranger Jack Parsell built the 30-foot by 40-foot Moose Creek Ranger Station log cabin that served as an office, cookhouse, and "honeymoon cabin" in 1921. "Jack broke in his new bride right that summer, cooking for the gang, tending telephone, milking the cow, etc., as all good rangers' wives did in the old days," noted 1921 summer assistant Bert Cramer. (Courtesy of the US Forest Service.)

By the time Parsell returned for his second assignment (1945–1955) there, Moose Creek Ranger Station had grown considerably under the four rangers who had succeeded him after 1922. An airstrip was constructed with "muscle and mule power" in 1931, and a barn, two residences, a warehouse, and other log buildings had been built by the time of this 1938 photograph. (Courtesy of the US Forest Service.)

The airstrip at Moose Creek Ranger Station became a development center for backcountry aviation where Forest Service history was made. Flying from there on July 12, 1940, pioneer smokejumpers Rufus Robinson (left) and Earl Cooley (right) made the first operational jump on a fire at Martin Creek. (Courtesy of the US Forest Service.)

There has been a Spotted Bear Ranger Station in the Flathead National Forest in Montana since 1906. These three log structures built between 1924 and 1934 by Victor "Big Vic" Holmlund of the Forest Service—a ranger's residence, an office, and a warehouse—comprise the Spotted Bear Ranger Station Historic District at the remote Spotted Bear Ranger Station. (Photograph by the author.)

Fifty-five miles south of Hungry Horse, Montana, Spotted Bear Ranger Station's 1985 log office building is the center of wilderness management operations in much of the more-than-one-million-acre Bob Marshall Wilderness. That famous wilderness is named for the Forest Service wilderness visionary and key pioneer of wilderness preservation. (Photograph by the author.)

The Gallatin National Forest's historic Madison Ranger Station, built in West Yellowstone, Montana, in 1924, included a ranger's residence (above) and an office building (below). Both these historic structures were moved to Dunbar Park in the town's historic district in 2010 for preservation and appropriate reuse. (Courtesy of the US Forest Service.)

Junior Smokejumpers, who completed a West Yellowstone Environmental Education Center program for youngsters, lined up for this "class picture" in front of the historic Madison Ranger Station office building now in the town's Dunbar Park and used for educational programs. (Courtesy of the National Smokejumper Center.)

A log cabin that began in 1921 as Boulder Creek Ranger Station in 1925 became Zion Creek Ranger Station and, in 1926, was dismantled and floated down the Lochsa River one mile and renamed Lochsa Ranger Station. As ranger Ralph Hand explained, the old Selway National Forest had too many ranger stations named for creeks. It is part of one of the West's premiere historic ranger stations. (Photograph by the author.)

Except for the old Boulder Creek Cabin, the log buildings of today's Lochsa Historical Ranger Station were built between 1928 and 1933 by the district rangers assigned at Lochsa and their crews. The ranger's dwelling (pictured) is beautifully preserved and furnished as it was in the 1930s, 1940s, and 1950s. (Photograph by the author.)

Lochsa was a remote ranger station until US Highway 12 reached it in 1956. It took four days to skid this 375-pound bathtub for the ranger's dwelling 12 miles up the Lochsa River Trail to the Lochsa Ranger Station construction site. (Courtesy of the US Forest Service.)

Mule strings such as this one being packed in front of the Lochsa Ranger Station combination building were the primary means of early transportation in the national forests. The combination building is beautifully preserved. Each of its four rooms is furnished and equipped for specific uses just as they were when the station was in use. (Courtesy of the US Forest Service.)

For almost a quarter century, from 1930 to 1953, the Forest Service's Ninemile Remount Depot—established five years before Ninemile Ranger Station at the same site—provided packers and saddle and pack animals for fighting fires and supporting backcountry work on national forest lands throughout the Northern Rockies. (Photograph by the author.)

The art of Forest Service mule packing was perfected at the Ninemile Remount Depot. The packer on the right is legendary forest ranger Bill Bell, made famous in Norman Maclean's memoir *USFS 1919: The Ranger, the Cook, and a Hole in the Sky*. (Courtesy of the US Forest Service.)

Forest Service mule packing—along with other aspects of Northern Region national forest history—is interpreted at the Ninemile Remount Depot visitor center. The ranger station and remount depot are about 25 miles west of Missoula, Montana, off Interstate Highway 90. (Photograph by the author.)

Ninemile Remount Depot mule strings were trucked to trailheads throughout the Northern Region. Today, the historic remount depot, a Ninemile Ranger District unit, is home to the Northern Region Pack Train, which supports wilderness management work and represents the Forest Service at rodeos, parades, and other events. (Courtesy of the US Forest Service.)

Half a million dollars was a big price to pay for a ranger station during the Great Depression—even one designed and built to accommodate two ranger districts. But that is what Uncle Sam spent on the Fenn Ranger Station, built on a bench above the Selway River by the CCC for the Nez Perce National Forest in Idaho between 1936 and 1939 as a headquarters for that forest's Middle Fork and Selway Ranger Districts. Construction began with an administration building (pictured), two warehouses, and two garages completed in 1937. By the end of 1939, a cookhouse, a gashouse, and two residences had been added. The barn was built in 1940. Given that Fenn Ranger Station remains in use, the government seems to have gotten the taxpayers' money's worth. With a new visitor center on the site, historic Fenn Ranger Station remains one of the Forest Service's showplace ranger stations, a monument to pioneer forest officer Maj. Frank A. Fenn, for whom it was named, and the men who built it. (Photograph by the author.)

The old Darby Ranger Station—called Darby Historic Ranger Station since the National Forest Centennial in 1991 and operated as a Bitterroot National Forest visitor center since—was completed in 1939 as the first ranger station in Darby, Montana. (Photograph by the author.)

The tool and tack room at Darby Historic Ranger Station, located at 712 US Highway 93 North in Darby, Montana, exhibits many tools of the early forest ranger's trade. (Photograph by the author.)

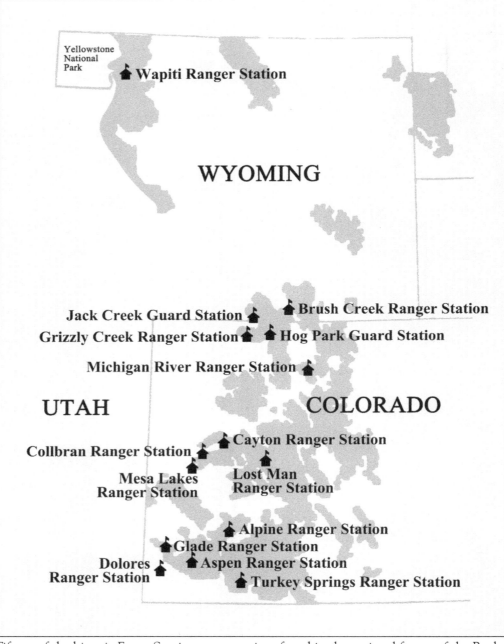

Fifteen of the historic Forest Service ranger stations found in the national forests of the Rocky Mountain Region, also called Region 2, and pictured in Chapter 2, are in the states of Colorado and Wyoming. (Map by Gary Asher and Les Joslin.)

Two
Rocky Mountain Region

The Rocky Mountain Region, a land of high peaks and vast forests of pine, fir, and spruce, is the birthplace of the National Forest System and the first forest ranger. Pres. Benjamin Harrison proclaimed the Yellowstone Timberland Reserve on March 30, 1891. Just seven years later, on August 8, 1898, young William R. "Bill" Kreutzer was appointed the first US forest ranger by the Department of the Interior. He transitioned to the new Forest Service in 1905. When Kreutzer (left) met with other early forest rangers at Mesa Lakes in the Battlement Forest Reserve in Colorado that year, they recognized the need for ranger stations. (Courtesy of the US Forest Service.)

America's first national forest claims one of America's first ranger stations. Shoshone National Forest, carved out of Yellowstone Timberland Reserve lands as the first unit of the National Forest System, is the site of historic Wapiti Ranger Station. When, in October 1903, ranger Harry Thurston rendezvoused with ranger Milt Benedict and his 16-packhorse loads of supplies at the confluence of the North Fork and the Elk Fork of the Shoshone River to construct a headquarters for the supervisor of the Shoshone division of the timberland reserve, the two rangers constructed the first ranger station built with government funds and still in service at its original location. In 1904, the first telephone line built by the US government for administration and protection of forest reserves linked Wapiti Ranger Station with the reserve's headquarters in Cody, Wyoming. Col. William F. "Buffalo Bill" Cody, for whom that city is named, and forester Gifford Pinchot were among the ranger station's earlier visitors. (Photograph by the author.)

Officially called Alpine Ranger Station at least through 1920, historic Alpine Guard Station consists of three log buildings—a residence, a barn, and a combination garage–equipment shed—for administration of Uncomphagre National Forest resources, particularly the grazing of as many as 90,000 sheep annually on high-country summer range. (Courtesy of the US Forest Service.)

In 2009, the Grand Mesa–Uncomphagre–Gunnison National Forests were awarded an American Restoration and Recovery Act of 2009 grant to restore the historic Alpine Ranger Station to provide public recreation opportunities. This restoration was completed by Colorado-based Historicorps, the Forest Service, and other partners. The station's elevation at 11,600 feet certainly merits the "Alpine" name. (Courtesy of Historicorps.)

Today's historic Cayton Ranger Station was called Johnson Spring Ranger Station in 1909 and 1910 when ranger James Grimshaw "Jim" Cayton and his bride—local rancher's daughter and schoolteacher Adelaide Dorothea "Birdie" Miller of Silt, Colorado—and other Forest Service rangers built the three-room log house and log barn from which Cayton ran his White River National Forest ranger district until 1919. During those years, the Caytons made Johnson Spring Ranger Station their "happy home." They fenced a half-acre yard around the house and seeded it with Kentucky bluegrass for a lawn. They transplanted small aspens to the yard, where they also planted memorial trees for three Forest Service men killed in France during World War I. The now-restored historic ranger station was renamed Cayton Ranger Station in his honor in 1939. (Courtesy of Mrs. James G. Cayton and White River National Forest.)

Jim Cayton and his wife, Birdie (left), lived at Cayton Ranger Station from 1909 to 1919 as Cayton (right) rode the range to manage the natural resources in his district. (Both, courtesy of Mrs. James G. Cayton and White River National Forest.)

Historic Cayton Ranger Station—restored through the efforts of the Cayton Ranger Station Foundation and listed in the National Register of Historic Places—remains to tell the story of the Caytons' years of service in the White River National Forest. (Photograph by the author.)

Authorized in 1907 to be built in the old Sierra Madre National Forest and built between 1910 and 1912 in the old Hayden National Forest, the log cabin and barn of the Hog Park Guard Station comprise the oldest remaining guard station in the Routt National Forest. Less than one-half mile south of the Colorado-Wyoming state line and a couple miles east of the Continental Divide, this remote historic guard station sits on the western bank of the Encampment River. Cutting railroad ties, an activity that peaked in the Hog Park area between 1900 and 1910 when some 300 "tie hacks" worked for a timber company later charged with timber trespass, seems to have stimulated establishment of the guard station. By the time the station was completed, the tie-cutting operation had moved on. Stabilized in 2007, historic Hog Park Guard Station is a national forest heritage asset with recreation potential. (Photograph by Midwest Archaeological Center, courtesy of the US Forest Service.)

The small log cabin once called Lost Man Ranger Station reflects the Forest Service's early appreciation of high-speed communications in national forest administration. The cabin, at an elevation of 10,500 feet and six miles west of 12,095-foot Independence Pass over the Continental Divide, was built in October 1913 in the White River National Forest by the Forest Service and the Mountain States Telephone Company primarily as a line cabin for crews who maintained telephone lines over the divide. Records show the Forest Service provided $85.04 in labor to the project while the telephone company supplied $50 worth of materials. Telephones were in use in the national forests since the earliest days of the Forest Service, which both built its own telephone lines and contracted with telephone companies. Rangers frequently took on the responsibility of maintaining the lines they used. By the late 1940s, Lost Man Guard Station was used only irregularly by Forest Service personnel as a temporary stopping place. (Courtesy of the US Forest Service.)

Glade Ranger Station was established in 1905 with the construction of a log cabin on a slight rise overlooking an area known as "the Glade" in what was then the Montezuma National Forest in southwestern Colorado. This 1916 wood-frame, four-room dwelling with a steeply-pitched pyramidal roof, built by W.E. Rittenhouse of Dolores, Colorado, replaced that cabin. Other structures are a barn of about the same age and a Depression-era garage, woodshed, and outhouse at about 8,400 feet and 18 miles from the nearest paved road in the northwestern reaches of the San Juan National Forest. This high and lonesome outpost, called Glade Guard Station after district ranger Cliff Chappell's headquarters moved into Dolores in 1935, survives as a five-building compound restored to mint condition by a privately funded three-year National Smokejumper Association effort coordinated by Forest Service heritage resource specialists. (Courtesy of Milt Griffith.)

CCC enrollees camped at Glade Ranger Station during the Great Depression when they built the garage, woodshed, and outhouse and otherwise improved the station from which Forest Service rangers administered timber sales and grazing permits. (Courtesy of the US Forest Service.)

The 1916 Glade Ranger Station dwelling looks as good as new following the historic station's 2008–2009 restoration by hard-working volunteer crews of former smokejumpers. Long painted brown, the buildings are a gleaming white again. (Courtesy of Julie Coleman, archaeologist, San Juan National Forest.)

Ranger George McClanahan completed construction of the Grizzly Creek Ranger Station log cabin—later called Grizzly Creek Guard Station—to Forest Service specifications in the Routt National Forest in Colorado in 1922. Instead of serving its original purpose as part of a proposed tree nursery and research facility, the cabin, along with a ranger's house and a barn, became the ranger station from which McClanahan administered his ranger district until 1929. The house was torn down in 1932, and the cabin became a guard station for one or two Forest Service employees. By 1961, the cabin's foundation was rotted and the roof leaked. But it was otherwise in good shape and remained in use until 1987. After that, it was overrun by critters and continued to deteriorate. Restoration of the cabin as a recreation and heritage resource began in 1999 and was completed in 2002. (Courtesy of the US Forest Service.)

George McClanahan and his family enjoyed living at the Grizzly Creek Ranger Station during the summer months he spent in his Routt National Forest ranger district until 1929, when his office was moved to Walden, Colorado. (Courtesy of the US Forest Service.)

Sixty-five years after ranger John Baird and his bride, Sally, married in Pagosa Springs, Colorado, they recalled their June 1929 honeymoon at Turkey Springs Ranger Station—and how the ranger station burned to the ground as they cooked the first breakfast of their new life together. The stove caught the roof on fire. Nearby sheepherders came to lend a hand. They helped rescue all the wedding presents and then the other objects from the cabin. Last out was the telephone Baird removed after calling forest supervisor Andrew Hutton to explain that the ranger station was burning down around him and that the supervisor's office would receive no more calls from him there. Not wanting to miss a free meal, the sheepherders even rescued the stove, with fire and breakfast still in it, and carried it to the lawn. All present had a picnic breakfast while the cabin burned. Baird rebuilt the Turkey Springs Ranger Station cabin (above) in 1930. (Courtesy of the National Park Service, Midwest Archaeological Center.)

In 1933, ranger Evan J. "Evie" Williams of the old Encampment Ranger District in the Medicine Bow National Forest in Wyoming began to build the Jack Creek Guard Station. With an allocation of $500, he purchased and delivered cement for the foundation and other building materials, and hired a crew to fell and haul logs. He completed the cabin with $75 of his own money and help from another ranger in 1934. It was not luxurious, but it did its job. Jack Creek Guard Station was used for timber, range, and wildlife management work for seven decades before it was made available as a recreation rental. Williams, who served his entire 35-year Forest Service career in the Medicine Bow National Forest, worked from the early years into modern times. He began in 1916 and retired in 1950. He accomplished many things, but historic Jack Creek Guard Station is his monument. His grave is about 20 yards north of the cabin. (Courtesy of the US Forest Service.)

Constructed by the CCC during the 1930s as a home and office for Forest Service rangers in the old Montezuma National Forest, the log and stone Aspen Ranger Station—now historic Aspen Guard Station in the San Juan National Forest—remains a remote reminder of how rangers lived and worked in the 1930s and 1940s. Crews continued to use the station through the mid-1970s, but the cabin fell into disuse and disrepair in the 1980s. After extensive restoration and rehabilitation work by Forest Service personnel and community partners, the historic log cabin was reborn in 1994 as the home of the Aspen Guard Station artist-in-residence program, which for many years hosted painters, writers, poets, photographers, and others selected to pursue their arts there. (Courtesy of the US Forest Service.)

In 1937, long before the streets in the western Colorado town of Collbran were paved, the Forest Service built Collbran Ranger Station as headquarters for a Grand Mesa National Forest ranger district. (Courtesy of the US Forest Service.)

As a result of district consolidation, the district ranger is now based in Grand Junction. In more recent years, historic Collbran Ranger Station has seen use as the Collbran office of the Grand Valley Ranger District, serving visitors and as a base for fire and project crews. (Courtesy of the US Forest Service.)

Since 1914, when a two-room log cabin and a log barn were built about 25 miles southeast of Walden, Colorado, to house a Forest Service officer to oversee a large timber operation in the Routt National Forest's old North Park Ranger District, there has been a Forest Service station along the upper reaches of the Michigan River. Over the years, this station has been variously known as Michigan River Ranger Station, Michigan Ranger Station, South Michigan Ranger Station, and—inexplicably—Michigan Creek Ranger Station before 1945, when it was redesignated and used as a guard station. After years as a guard station, it was used off and on as seasonal housing until 1998. A fine example of Rocky Mountain Region log architecture, it retained potential as both Forest Service housing and a recreation rental cabin. (Courtesy of the US Forest Service.)

The four adobe-style buildings of the historic Delores Ranger Station, built in Dolores, Colorado, by the CCC between 1937 and 1940 as headquarters for the Delores and Glade Ranger Districts of the Montezuma National Forest, blend well with the surrounding sandstone cliffs. After the Montezuma was incorporated into the San Juan National Forest in 1947, the station became headquarters of that forest's Delores Ranger District, which incorporated the old Glade and Rico Ranger Districts. (Courtesy of the US Forest Service.)

Delores Ranger Station, pictured in 2005, is no more. Consolidation of the Delores and Mancos Ranger Districts closed the station in September 2006, and in June 2011, the Forest Service turned the historic structures over to the Delores Pubic School District for school meetings and other educational uses. (Photograph by the author.)

There has been a ranger station at the Mesa Lakes on the Grand Mesa since at least 1905, when pioneer rangers like Bill Kreutzer and Jim Cayton rangered western Colorado's forest reserves for the brand-new Forest Service. Ranger Walt Anderson packs hay from the second set of ranger station buildings there in 1924. (Photograph by Ray Peck, courtesy of the US Forest Service.)

Mesa Lakes Ranger Station was reconstructed in 1939 and 1940. This June 1, 1940, photograph shows the construction of a new log ranger's residence (left) and a new log office (right) around the 1922 ranger station log cabin (center). After the new buildings were finished, the 1922 building was demolished. (Photograph by H.P. Gaylor, courtesy of the US Forest Service.)

Mesa Lakes Ranger Station residence (above left) and office (above right) appear during the 1940s, when it was a ranger district headquarters, just about as it did more recently (below) when it housed seasonal Forest Service personnel and volunteers who provided visitor information during the summers. The opening of a new Grand Mesa Visitor Center on the Grand Mesa National Scenic Byway replaced the historic ranger station's visitor service function, and it became a recreation rental. (Above, photograph by Higgins, courtesy of the US Forest Service; below, photograph by the author.)

High in southern Wyoming, in a grassy clearing where the dense lodgepole pine forest of the Snowy Range's western slope yields to the open sagebrush of the Kindt Basin, historic Brush Creek Ranger Station—three log structures built between 1937 and 1941 to replace an older ranger station—has been retained in service as a Medicine Bow–Routt National Forest work center and visitor center. The first Brush Creek Ranger Station, built about a mile to the northwest in 1905 for a Medicine Bow Forest Reserve ranger, was originally called Drinkhard Ranger Station. But that name, not deemed proper for the new Forest Service's image, was changed to Brush Creek in 1914. The three log buildings that currently comprise the station were built by the CCC. (Courtesy of the US Forest Service.)

Three

SOUTHWESTERN REGION

The national forests of the Southwestern Region, or Region 3, include over 20 million acres in the states of Arizona and New Mexico. These lands—generally the higher, cooler, and wetter lands of this arid region—provide most of the two states' usable water as well as timber for their sawmills, grazing for livestock, habitat for wildlife, and myriad other benefits including a variety of outdoor recreation opportunities. In the early years, Southwestern Region forest rangers commonly used abandoned ranch and mine cabins for stations. (Map by Gary Asher and Les Joslin.)

Built in 1906 in the old Grand Canyon Forest Reserve, historic Jump Up Ranger Station is the oldest existing Forest Service ranger station on the Kaibab Plateau of northern Arizona. Most of the plateau was set aside in 1893 by Pres. Benjamin Harrison as part of that forest reserve of which the Forest Service assumed administration in 1905. The forest reserve land north of the Grand Canyon was named Kaibab National Forest in 1908, and that south of the canyon became Tusayan National Forest in 1910. Administration of Grand Canyon National Park, established between the two national forests in 1919, went to the National Park Service established in 1916. During those early years and the ensuing decades, Jump Up Ranger Station served as a Forest Service ranger station and as temporary quarters for cattle ranchers. Restored in 2009, the historic cabin now shelters forest visitors. (Courtesy of the US Forest Service.)

Hull Tank Ranger Station, shown about 1925, was built in 1888 by sheep ranchers Phillip and William Hull and acquired by the Forest Service in 1907. In 1910, with the establishment of the Tusayan National Forest, it became an important administrative site. Rangers there were responsible for a large area that, until the establishment of Grand Canyon National Park under National Park Service administration in 1919, included part of the Grand Canyon. (Courtesy of the US Forest Service.)

Historic Hull Tank Ranger Station, now within the Coconino National Forest, was stabilized in 1989 and 1990. Using original methods and materials, Forest Service personnel from throughout the West helped replace rotten logs in the main cabin and barn, reroofed both cabins, rebuilt the cracked chimney in the main cabin, and replaced clapboards on the barn gable end. Continuing maintenance and restoration will keep Hull Cabin fit for use as a recreation rental. (Photograph by the author.)

About 40 miles north of the Grand Canyon, high on the Kaibab Plateau within the Kaibab National Forest, Forest Service ranger Will Mace (leaning against the porch post) completed construction of Jacob Lake Ranger Station in 1910. The cabin and barn face the meadow that surrounds Jacob Lake, a natural pool named for Mormon pioneer Jacob Lake and converted into a cattle tank. (Courtesy of the US Forest Service.)

Jacob Lake Ranger Station was built about 20 years after cattle grazing and lumbering, which began on the Kaibab Plateau about 1890, demonstrated the need for government management of these natural resources. After many decades of use, the historic Jacob Lake Ranger Station cabin has been restored to its original appearance to evoke the life and work of Will Mace and other early Forest Service rangers. (Photograph by the author.)

This log cabin was built in 1908 by rancher Charles Babbit of Flagstaff—grandfather of future governor of Arizona and Secretary of the Interior Bruce Babbit—and Bill Dickinson of Sedona. Both their families grazed cattle on Coconino National Forest rangelands. The Forest Service acquired the cabin in 1912 and named it Woodland Ranger Station; by 1917, when this photograph was taken, it was renamed Apache Maid Ranger Station. (Courtesy of the US Forest Service.)

Ranger James Bailey valued the cabin as living quarters for the nearby Apache Maid Lookout, and before long, the cabin adopted the lookout's name. After decades of use, historic Apache Maid Ranger Station in the Coconino National Forest was restored and, in May 2010, was added to the Southwestern region's popular Room With a View cabin rental program. (Courtesy of the US Forest Service.)

Aldo Leopold, a young Yale Forest School graduate who served as Carson National Forest supervisor from May 1911 until March 1913 and later became a leading American conservationist, needed a residence for himself and his bride—the former Estella Bergere of Santa Fe—at his Tres Piedras, New Mexico, headquarters. (Photograph by John D. Guthrie, courtesy of the US Forest Service.)

Ranger Walter J. Perry, pictured with his family in front of the house the Leopolds called "Mia Casita" about 1916, supervised construction of the house. An office, built about the same time as the house but six miles west as Cow Creek Ranger Station, was moved to Tres Piedras about 1917, and a barn and several small outbuildings were erected on the site. (Courtesy of the Walt Perry collection.)

The American flag flies above the office and the Leopold House at Carson National Forest headquarters in Tres Piedras, New Mexico. The Ford truck parked outside the house was acquired in 1919. Tres Piedras is named for the large granite formations that dominate the site. (Courtesy of the US Forest Service.)

The restored Leopold House at the Old Tres Piedras Administrative Site, Carson National Forest, is a magnificent memorial to the early days of the Forest Service, to Aldo Leopold's pioneering land ethic, and to Walt Perry's ability to do anything he was asked. Carolyn D. (Wilton) Bird, who took this photograph, lived there as a child. (Photograph by Carolyn D. Bird.)

The historic Portal Ranger Station office and quarters building, near the mouth of Cave Creek in the Chiricahua Mountains of southeastern Arizona, was built by the CCC in the Coronado National Forest in 1933. By that time, the Southwestern Region had developed standard building plans, and the regional office in Albuquerque, New Mexico, suggested use of either the standard bungalow style or the Pueblo style. Forest supervisor Fred Wynn proposed a modified version of the bungalow-style plan that used locally abundant cobblestone for the walls. (Photograph by J.J. Lamb, courtesy of University of Arizona School of Natural Resources and the Environment.)

When, in May 1909, a forest guard—later ranger—named Henry Woodrow made camp, he did not know he had located his summer base for the next 33 years. The ranger stations built there in 1912 and 1933, at the confluence of White Creek and the West Fork of the Gila River in the Mogollon Mountains of southwestern New Mexico, served as summer headquarters of the Gila National Forest's old McKenna Park Ranger District for 59 years. In 1924, as a result of Forest Service officer Aldo Leopold's wilderness advocacy, Woodrow's district became part of the Gila Wilderness Area administratively designated by district forester Frank C.W. Pooler, the first area in the National Forest System so designated. After the passage of the Wilderness Act of 1964 and Congress's designation of the Gila Wilderness, wilderness ranger operations were based at historic White Creek Ranger Station. (Courtesy of the US Forest Service.)

Big Springs Ranger Station, at the base of a limestone cliff on the Kaibab Plateau north of the Grand Canyon, is named for the source of spring water that cascades down that cliff and flows into two ponds on the site set aside in 1908 for a ranger station. The current office, dwelling, and barn built in 1934 had a predecessor in a ranger's residence built soon after the site was acquired and used for two decades. The station grew when the housing boom after World War II and the advent of logging trucks led to large-scale logging operations in the Kaibab National Forest. Big Springs Ranger Station was headquarters of the Big Springs Ranger District until 1974, when it was merged with the Jacob Lake Ranger District to form the North Kaibab Ranger District headquartered in Fredonia, Arizona. It remains a Forest Service administrative site. (Courtesy of the US Forest Service.)

About a dozen miles northeast of downtown Tucson, Arizona, this Coronado National Forest portal sign directs visitors to Lowell Ranger Station; in 1938, it was the headquarters of the Santa Catalina Ranger District and the main visitor contact point for the popular Sabino Canyon Recreation Area until the adjacent Sabino Canyon Visitor Center opened in the 1960s. (Photograph by C.C. Cunningham, courtesy of the US Forest Service.)

Historic Lowell Ranger Station's three Pueblo-style buildings—an office, a residence, and a garage-shop building constructed by the CCC—look right at home among the saguaro, cholla, and ocotillo of their Sonoran Desert setting. Still in use, these structures house Forest Service offices. (Photograph by Richard M. Lewis, courtesy of the US Forest Service.)

High in the Chiricahua Mountains of southeastern Arizona, in a narrow canyon in the Chiricahua Wilderness, the log buildings of the Coronado National Forest's remote Cima Park Fire Guard Station look just about as they did when the CCC completed them in 1934. The cabin was home to a fire dispatcher, connected with fire lookouts by telephone, and a seasonal fire crew. (Courtesy of the US Forest Service.)

Although modern firefighting techniques have made the Cima Park Fire Guard Station headquarters role a thing of the past, the station's historic cabin and barn continue to be used by wilderness rangers, trail crews, and fire crews as they double as a heritage resource. (Courtesy of Charmane Powers.)

A series of "make-do" Crown King Ranger Station facilities served as headquarters of the first national forest ranger district in Arizona for 30 district rangers over almost eight decades. High in the rugged Bradshaw Mountains southeast of Prescott, both the ranger station and the community were named for the 1880s Crown King Mine. By the time the Prescott Forest Reserve was established in 1898 and what became Crown King Ranger District was set up in 1902, much of the area's ponderosa pine forest had been cut to provide mine timbers and fuelwood for the burgeoning mines. As a result, pioneer ranger Ed Ancona's early-20th-century work from his personally purchased "ranger station" emphasized small sales of mining timbers and a lot of range management and trail work. He got on well with tough fellows, those cattlemen and cowboys of that day with whom he rode on roundups and even got to do some trail work. A couple decades after Ancona's departure, the Prescott National Forest's historic Crown King Ranger Station office, seen here, was built in 1936. (Courtesy of the US Forest Service.)

Also built by the CCC in 1936 were the Crown King Ranger Station ranger's dwelling (above) and barn-garage (below), which, along with the 1936 office building, remain on the site. Sixteen district rangers administered the many uses of the 161,950-acre Crown King Ranger District from the office and these buildings from the time they were built until 1979, when the district was combined with another to form the Prescott National Forest's larger Bradshaw Ranger District. Except for minor changes, these historic structures appear just about as they did when built and remain in service. (Both photographs by R.H. Lewis, courtesy of the US Forest Service.)

Historic Moqui Ranger Station is an intact compound of six Forest Service structures built between 1939 and 1942 by the CCC and now part of the Tusayan Ranger Station in the Kaibab National Forest just south of the main entrance to Grand Canyon National Park. Set among magnificent ponderosa pines, Moqui Ranger Station's buildings of native sandstone and wood—an office, a dwelling, a barn, a garage, and a seed house—were and remain a place of beauty. Above, a visitor and district ranger Clyde Moose converse in front of the Moqui Ranger Station office in 1948. The ranger's dwelling (below) afforded comfortable living. (Above, photograph by E.L. Perry; below, photograph by W.G. Mann, both courtesy of the US Forest Service.)

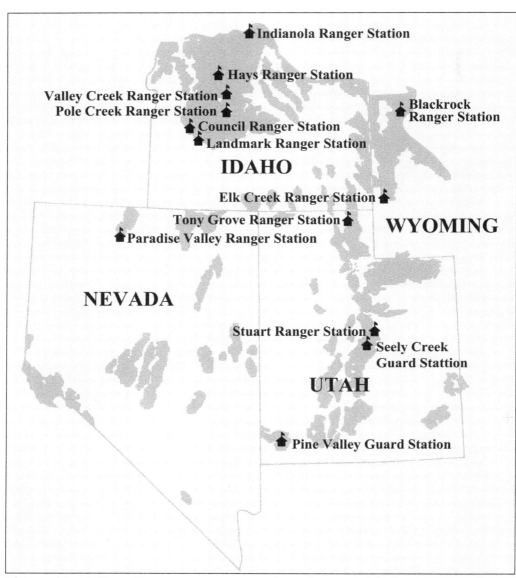

Thirteen historic Forest Service ranger stations and guard stations of the Intermountain Region, or Region 4, profiled in this chapter are located in national forests in southern Idaho, Nevada, Utah, and western Wyoming. (Map by Gary Asher and Les Joslin.)

Four

INTERMOUNTAIN REGION

The national forests of the Intermountain Region comprise 34 million acres of National Forest System lands within Utah, Nevada, western Wyoming, southern Idaho, and a small part of eastern California. This region, as did the rest of the West, challenged early Forest Service rangers—such as this ranger on patrol in Logan Canyon, Utah—who, after long periods of work in the field, found respite in their ranger station offices and residences. (Courtesy of the US Forest Service.)

An amazing Austrian immigrant named Rudolph "Rosie" Rosencrans, who became the first ranger of the Buffalo Ranger District in what is now the Bridger-Teton National Forest in northwestern Wyoming, is remembered at the Rosencrans Cabin Historic District. There, the one-room log Blackrock Ranger Station office (pictured) he and fellow ranger John Alsop built in 1904 and a larger log residence—known as "Rosie's Office" and "Rosie's Cabin," respectively—represent his remarkable Forest Service career. Born in 1875, this technically trained son of Austria's chief forester left the Austrian navy in the late 1890s for Wyoming, became an American citizen and, in 1904, the first US Department of the Interior forest ranger for the then Teton Division of the Yellowstone Timberland Reserve. In 1905, he became a Forest Service ranger when forest reserve administration was transferred to that new agency. In 1908, the Teton Division became the Teton National Forest. (Photograph by the author.)

Rosencrans personified Gifford Pinchot's ideal of the educated, dedicated, outdoors-wise professionals needed to staff the young Forest Service. He impressed such visitors as Pres. Theodore Roosevelt and his boyhood hero Col. William F. "Buffalo Bill" Cody, whom he hosted. He did it all, from fighting forest fires to building trails to rescuing the lost and recovering the dead. He won ranchers over to Forest Service methods of grazing the range and produced the first maps of his district. Rosie was the consummate forest ranger, and everyone knew it. Failing eyesight—perhaps a result of drawing intricate maps such as the one below of the Blackrock Ranger Station vicinity in poor light, or the glare of sun and snow, or both—forced him to retire in 1928 at age 55. He died in Jackson, Wyoming, in 1970 at age 94. (Both, courtesy of the US Forest Service.)

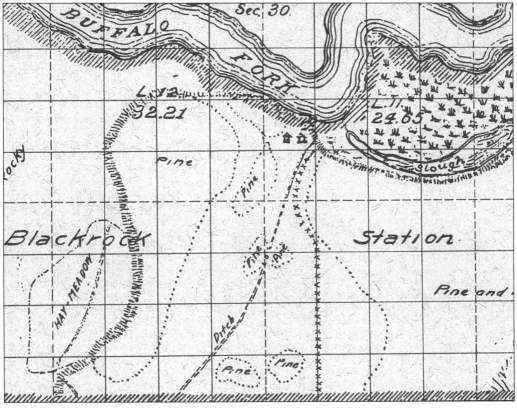

The historic Seely Creek Guard Station was built in a beautiful Wasatch Plateau wildflower meadow for Utah's old Manti National Forest in 1907 and 1908. Local men using materials bought in nearby Ephraim constructed the house, which remains on the site, and a large stable as summer headquarters for the Forest Service ranger in that part of the forest. With the availability of Depression funding, the CCC improved the station in 1934. As time passed and transportation improved, Seely Creek Guard Station use declined. By the late 1990s, it was in poor condition and slated for removal when Forest Service personnel and local community members moved to save it for public recreation use as part of the Manti–La Sal National Forest's cabin rental program. (Courtesy of the US Forest Service.)

A Forest Service ranger uses the telephone on the front porch of Tony Grove Ranger Station in Cache National Forest, Utah, in 1937. Built in 1908 in Logan Canyon, some 20 miles northeast of Logan, the station has been in continuous use for a wide range of purposes ever since. (Courtesy of the US Forest Service.)

The 1908 Tony Grove Ranger Station log cabin remains the centerpiece of the Tony Grove Ranger Station Historic District along the Logan Canyon National Scenic Byway in today's Uinta-Wasatch-Cache National Forest. (Photograph by the author.)

Indianola Ranger Station was headquarters of the old Indianola Ranger District in Salmon National Forest, Idaho, from 1908 to 1972. After those years, the ranger station became Indianola Forest Service Station in the Salmon-Challis National Forest and a base for a helicopter attack ("helitack" in wildland firefighter parlance) fire crew. (Courtesy of the US Forest Service.)

This memorial for Indianola helitack crew members Jeff Allen and Shane Heath, lost in the Cramer Fire near their Indianola Forest Service Station base on July 22, 2003, is a poignant reminder of the dangers faced by wildland firefighters. (Photograph by the author.)

Seemingly lost in the blue-green vastness of the Sawtooth Valley of Idaho, historic Pole Creek Ranger Station—later called Pole Creek Guard Station—built by ranger William H. "Bill" Horton in 1908 and 1909, is the oldest Forest Service–constructed building in the Sawtooth National Forest. (Photograph by the author.)

Horton, who served at Pole Creek every summer from 1908 to 1929, welcomed forest users and visitors to the Pole Creek Ranger Station he built and staffed. After the new Valley Creek Ranger Station was constructed in 1933 (see pages 74 and 75), seasonal Forest Service personnel worked from the Pole Creek station until the 1950s, when it was abandoned. (Photograph by Milton S. Benedict, courtesy of the US Forest Service.)

Fires that raged in the Payette National Forest in Idaho in 1994 almost claimed the historic Hays Ranger Station, a log cabin built by the Forest Service in 1913 that served as a Warren Ranger District station in the old Idaho National Forest until 1918. But firefighters foamed and saved the cabin, which remains the oldest standing Forest Service building in the Payette created in 1944 by combining the Idaho and Weiser National Forests. (Courtesy of the US Forest Service.)

A small log cabin, built in 1914 and known as Elk Creek Ranger Station and later as Elk Creek Guard Station, remains the oldest surviving Forest Service administrative building still in its original location in the Bridger-Teton National Forest in Wyoming. Typical of Forest Service stations built in the early 1900s, it was used by rangers monitoring livestock grazing and timber cutting and controlling forest fires. (Courtesy of the US Forest Service.)

Landmark Ranger Station, established in 1924 and, for years, the summer headquarters of Idaho's "old" Payette National Forest before the station was transferred to the Boise National Forest in 1944, was closed and all but abandoned in the late 1990s. (Photograph by Shipp, courtesy of the US Forest Service.)

The picturesque Landmark Ranger Station compound of log structures and frame structures with log siding, ideally situated to manage the vast region district rangers based there oversaw, faced an uncertain future until Cascade District ranger Carol McCoy-Brown arrived in 2005. Recognizing its importance and historical significance, she set about bringing it up to date for continued use while maintaining its historical and architectural integrity. (Photograph by Shipp, courtesy of the US Forest Service.)

The original Valley Creek Ranger Station in Stanley, Idaho, near the confluence of Valley Creek and the Salmon River, was a three-room log house completed in 1909 while either Edgar P. Huffman or Wallin Job was district ranger—official records and old-timer recollections differ on this point. Transferred from the Sawtooth National Forest to the Challis National Forest in 1913, it served as headquarters for several Stanley Basin Ranger District rangers until 1933. (Courtesy of the US Forest Service.)

In 1932 and 1933, during ranger Merle Markle's tenure, a new Valley Creek Ranger Station office was built. Markle cut and hauled the logs for the new building, and his wife, Kathleen, helped with the peeling. Establishment of the CCC in April 1933 provided labor for the actual construction of the beautiful log structure. Additional buildings rounded out the new ranger station compound by 1936. (Photograph by the author.)

CCC enrollees from the Redfish Lake Camp helped complete the Valley Creek Ranger Station main building in 1933. This CCC worker takes a break to chat with a young woman. (Photograph by Kenneth D. Swan, courtesy of the US Forest Service.)

Valley Creek Ranger Station was a busy district headquarters for decades. When the Sawtooth National Recreation Area was created in 1972, the station was transferred back to the Sawtooth National Forest but was no longer needed as a ranger district headquarters. In the late 1980s, the Sawtooth Interpretive and Historical Association assumed care of the historic station and operated a pioneer museum in the main building. (Courtesy of the US Forest Service.)

Historic Council Ranger Station, in the northern Weiser Valley near the base of Council Mountain and within the city limits of Council, Idaho, was built between 1933 and 1936 by the CCC as headquarters for one of the three ranger districts in the old Weiser National Forest. In 1941, the Weiser and Idaho National Forests were combined to form the Payette National Forest. (Courtesy of the US Forest Service.)

District ranger Dewitt Russell assists a Weiser National Forest visitor on September 15, 1941. After a new district ranger's office was built, this historic office building was converted to a visitor center, and the other four frame buildings erected on the site between 1933 and 1936 remained in use. (Courtesy of the US Forest Service.)

CCC crews constructed the Paradise Valley Ranger Station compound comprising an office, residence, garage-storeroom, barn, and water system, which included a pump house and large stone cistern, between 1934 and 1941 for the Santa Rosa Ranger District in Nevada. Transferred back and forth between the Humboldt and Toiyabe National Forests, the district was ultimately attached to the Humboldt National Forest. (Photograph by W.H. Shaffer, courtesy of the US Forest Service.)

The Paradise Valley Ranger Station bunkhouse is a converted barn moved to Paradise Valley in 1938. Converted to a work center, the historic Paradise Valley Ranger Station compound in today's Humboldt-Toiyabe National Forest is a collection of eight historic structures, which, in white paint with green trim, look just about as they did when completed just before World War II. (Courtesy of the US Forest Service.)

Historic Stuart Ranger Station—now also called Stuart Guard Station and operated as a Manti–La Sal National Forest visitor center along the Huntington and Eccles Canyons National Scenic Byway in central Utah—was built by the CCC in 1934. It was named for chief forester Robert Y. Stuart, who died in office on October 23, 1933. It was restored between 1997 and 1999. (Photograph by the author.)

Stuart Ranger Station exhibits tell the stories of the Manti–La Sal National Forest and the roles of the Forest Service and surrounding communities in managing its natural resources. The historic station is dedicated to the young men of the CCC. (Photograph by the author.)

The CCC constructed the Pine Valley Guard Station in 1935 to accommodate the Forest Service recreation guard in charge of the adjacent campgrounds and other facilities in the Pine Valley Recreation Area in the Dixie National Forest about 30 miles north of St. George, Utah. After more than five decades, use and maintenance dropped off, and by 1998, the cabin was in poor condition. But the Dixie National Forest saw potential for the cabin as a recreation rental and had a restoration plan ready to implement when the American Recovery and Reinvestment Act of 2009 funded "shovel ready" projects. Contractors beautifully restored the cabin from foundation to roof, installing everything needed—including updated plumbing and electricity—to make it comfortable, safe, and ready to rent to the public beginning in 2010. (Photograph by Marian Jacklin, courtesy of the US Forest Service.)

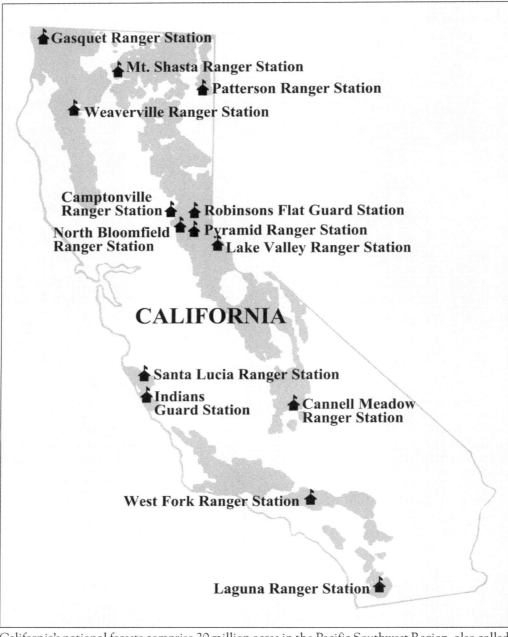

California's national forests comprise 20 million acres in the Pacific Southwest Region, also called Region 5 and once called the California Region. These forests, like the state itself, are lands of contrast. Wrapped around the Great Central Valley, they occupy the higher elevations of the Coast Ranges, the Sierra Nevada, and the Cascade Range—and they harbor many historic ranger stations. (Map by Gary Asher and Les Joslin.)

Five
PACIFIC SOUTHWEST REGION

Habitats within California's national forests range from the chaparral, oak and pine woodlands—such as surrounded the old Sierra National Forest headquarters in North Fork, pictured here, which burned in 1992—and forests of coastal redwoods, the Douglas fir of the northwest, and the yellow pines of the Sierra—that provide vital water as well as timber, grazing, recreation, abundant natural beauty, and other benefits to the nation's most populous state. (Courtesy of the US Forest Service.)

Rangers who patrolled California's first forest reserve, the San Gabriel Timberland Reserve set aside by Pres. Benjamin Harrison in 1892, built California's first ranger station on the West Fork of the San Gabriel River in 1900. Reconstructed in 1983 at the nearby Chilao Visitor Center on the scenic Angeles Crest Highway, historic West Fork Ranger Station gives Angeles National Forest visitors a glimpse of early forest ranger life. (Photograph by the author.)

Ranger Louie Newcombe (far left), who supervised construction of West Fork Ranger Station for $70 in federal funds and stayed on as ranger in charge for three years, and (from left to right) rangers Phil Begue, Willard Sevier, and Jack Baldwin fought fires, built trails, helped lost hikers and fishermen, and helped build two more San Gabriel Forest Reserve ranger stations in 1902 and 1903. (Courtesy of Big Santa Anita Historical Association.)

The historic Cannell Meadow Ranger Station, now called Cannell Cabin, was the first Forest Service ranger station on the Sierra Forest Reserve from which today's Sequoia National Forest was carved. This log cabin was built in 1905 and 1906 by Oswald Homer "Walt" Klein, the second forest ranger in the Kern Valley but the first of Gifford Pinchot's new Forest Service, and a couple others. Klein and his successors made the Cannell Meadows Ranger Station the summer headquarters of the Cannell Meadow Ranger District, later based in Kernville, during the early years in which the Forest Service sorted out its administration of the lands it assumed from the General Land Office. Over the years, the Forest Service put the renamed Cannell Meadow Guard Station to many uses, including fire patrols based there until 1983. Restoration of the cabin began in 1996 with a Passport in Time project planned and led by Boise National Forest–based historic building restoration specialist Joe Gallagher. Loren Ross, the last Cannell Meadow fire guard, volunteered on that project. (Photograph by the author.)

Sited on Santa Lucia Creek, Santa Lucia Ranger Station was constructed of adobe bricks and locally-collected river cobbles in 1908, two years after the establishment of the Monterey Forest Reserve in 1906, which was renamed the Monterey National Forest in 1907, incorporated into the Santa Barbara National Forest in 1919 and, in turn, renamed Los Padres National Forest in 1936. (Courtesy of Jeffrey Whitmore.)

Santa Lucia Ranger Station housed Forest Service rangers who patrolled on foot and horseback to protect the forest from wildfire, illegal grazing, wildlife poaching, and timber theft. After its ranger station days, the Girl Scouts operated a summer camp there from the 1930s into the 1980s. Wildfires blackened the surrounding mountains in 2008, and protective barriers were installed around the historic adobe to prevent mudslide and debris flow damage. (Photograph by the author.)

The historic Pyramid Ranger Station building, constructed in 1910 and 1911 along the American River between Placerville and Lake Tahoe, is the oldest standing Forest Service structure in the Eldorado National Forest. Forest rangers and fire guards carried out their duties from this summer station along US Highway 50, the main route between San Francisco and Lake Tahoe, for at least the next 60 years. (Courtesy of the US Forest Service.)

From 1993 to 2005, the Pyramid Ranger Station building was sited at Owens Camp Fire Station, about 10 miles from its original site, where the Forest Service fire crew rehabilitated it to serve as the office. When that fire station closed, the historic building was moved about 30 miles to Camino and located in front of the Placerville Ranger Station where it opened in 2008 as an Eldorado National Forest visitor center. (Courtesy of the US Forest Service.)

The log cabin that was Laguna Ranger Station, the "first permanent ranger station in the Laguna Mountains," according to the sign that bears its current El Prado Cabin name, still stands on the Cleveland National Forest slope overlooking *el prado*—"the meadow" in Spanish—where Forest Service ranger Carl Brenner, his assistant, and some local men built it in a hurry but well in 1911. Its construction was inspired by a San Francisco district (now regional) office suggestion that it was needed there—a need proved by growth in summer recreation during the 1920s. By 1925, to support increased visitor use of the Laguna Mountain Recreation Area, a log campground registration booth had been erected in front of the cabin. This helped preserve one of California's few remaining early Forest Service log cabins. As early as 1967, it was maintained as what the forest supervisor termed "a feature attraction for recreation sightseers." Extensive restoration in 1989 by members of the Laguna Mountain Volunteer Association ensured El Prado Cabin would remain at the edge of its meadow for many years. (Photograph by the author.)

The first Forest Service rangers who served in the Lake Tahoe Forest Reserve—which, combined with some other lands, became the Tahoe National Forest in 1907—had no ranger station within the Tahoe Basin from which to work. Finally, in 1911, funds were allocated, and in 1912, the Lake Valley Ranger Station house shown in this 1918 photograph and a barn were completed. When in 1920 a larger ranger station was built at Meyers, three miles away, the 1912 house was rented to a private citizen for almost 40 years. After that lease was relinquished in 1959, the Forest Service used the facility as a seasonal pack station. In 2004, the pack station that was the original Lake Valley Ranger Station was renamed Hawksworth Pack Station in honor of Forest Service packer James Hawksworth and his wife, Donna, who packed out of the station for 30 years. (Courtesy of the US Forest Service.)

Robinsons Flat Guard Station, built between 1913 and 1916 in the Tahoe National Forest, is among the older remaining Forest Service guard stations in California. An important base for fire patrol and control, it has been maintained in good condition for over a century. (Courtesy of the US Forest Service.)

Just across Humbug Creek from the historic Gold Rush town of North Bloomfield—centerpiece of Malakoff Diggins State Historic Park—old North Bloomfield Ranger Station spans the history of the Tahoe National Forest from the early years of the Forest Service through World War II. (Photograph by the author.)

During the early 1990s, Passport in Time volunteers helped the Tahoe National Forest restore historic North Bloomfield Ranger Station, which was built on a site the Forest Service purchased from the town of North Bloomfield in 1909. The ranger station was headquarters of the old North Bloomfield Ranger District until 1946. (Courtesy of the US Forest Service.)

During the summers of 1920 and 1921, Ranger Ben L. Johnson of the Modoc National Forest's old South Warner Ranger District built a one-dwelling ranger station in the southern Warner Mountains of northeastern California. It was the Modoc National Forest's first structure based on district forester Coert DuBois's standard plans and is the forest's only surviving pre-1930s building. The station overlooked Patterson Meadow, named for the Patterson family, who settled the area in 1905 and operated a nearby sawmill. The station is now called Patterson Guard Station. During its first dozen years, it saw only occasional use. After 1934, when the CCC added a garage, Patterson Guard Station housed summer fire crews. Regular maintenance has preserved the buildings in good condition. Foaming by firefighters saved the historic station from the 34,000-acre Blue Fire of 2001. (Photograph by the author.)

Indians Guard Station, built in 1929 in an area of the old Santa Barbara National Forest known as "the Indians," provided a home for forest rangers who patrolled the land and fought fires. During the Great Depression of the 1930s, CCC units stationed at Indians constructed the road along Arroyo Seco River. Restored in the mid-1990s and retained for its heritage values, historic Indians Guard Station allows Los Padres National Forest visitors a glimpse of Forest Service life in the 1930s and 1940s. (Courtesy of the US Forest Service.)

The historic part of the Gasquet Ranger Station compound, the Gasquet Ranger Station Historic District, in the Six Rivers National Forest in northwestern California between US Highway 199 and the Middle Fork of the Smith River, was built by the CCC between 1933 and 1939. The site was then in the Siskiyou National Forest and part of the Pacific Northwest Region of the National Forest System. (Photograph by Bob Foley.)

All but one of this historic station's buildings reflect the Pacific Northwest Region's standard Cascadian Rustic architecture of the time. At the insistence of district ranger Adolph Nilsson's wife, however, the ranger's residence was built in the Colonial Revival style popularized by the reconstruction of Colonial Williamsburg, Virginia, in the 1920s, from a plan in the July 1930 *Ladies Home Journal*. (Photograph by the author.)

Construction of historic Camptonville Ranger Station—still in use as a Forest Service fire station until the early 21st century—was just one of many changes wrought by Frank W. Meggers during his 1927 to 1945 service as district ranger in the Tahoe National Forest's old Camptonville Ranger District. As retired ranger Meggers put it in 1975, "They came along and said, 'If you had all the money you wanted, what could you do to help your district?'" His answer was a headquarters to replace his rented office over the town's saloon. He got it when the CCC built the Camptonville Ranger Station compound—an office, two residences, and two garages—to the California Region's standard designs between 1934 and 1936. It housed the district operations until 1971. In 1959, Camptonville Ranger Station was the command post of the pitched battle during which a thousand firefighters barely saved the town and the ranger station from a major forest fire. By the early 21st century, saving Camptonville Ranger Station for the future appeared a more formidable challenge. (Photograph by the author.)

The historic Mount Shasta Ranger Station, a compound of 14 buildings and three radio towers, was built between 1934 and 1940—to standard California Region plans with Depression-era Emergency Conservation Work funds by local experienced men and CCC labor—in the town of Mount Shasta to serve as the headquarters of Shasta National Forest. (Courtesy of the US Forest Service.)

The compound served as the Shasta National Forest headquarters for 20 years. It has been the Mount Shasta Ranger Station, headquarters of the Mount Shasta Ranger District, since 1954 when the Shasta and Trinity National Forests were combined as the Shasta-Trinity National Forest headquartered in Redding, California. (Photograph by the author.)

The historic Weaverville Ranger Station is the architectural and historical twin of historic Mount Shasta Ranger Station. Fifty miles southwest of the town of Mount Shasta, it was a 10-building compound begun in 1934 as the headquarters of Trinity National Forest, which was combined with Shasta National Forest in 1954 to form the Shasta-Trinity National Forest. The site on which today's Weaverville Ranger Station compound was built, at the confluence of three streams flowing out of the Trinity Alps into the Trinity River, was extensively placer-mined for gold in the late 19th century. Sometime after World War I, probably in the early 1930s, the Forest Service acquired this property, then a jumble of town lots and mining claims, on which to build the Trinity National Forest headquarters, which later became the Weaverville Ranger Station. As impressive as the buildings are the grounds. A CCC crew stabilized the three stream channels and carried out a landscape plan shortly after the buildings were completed. This landscape has matured beautifully over the decades. (Photograph by the author.)

National forest lands totaling more than 24 million acres make up the Pacific Northwest Region, or Region 6. Bisected north to south by the Cascade Range, the region's wet western side and dry eastern side contain diverse terrain, climate, and vegetation—and many historic ranger stations. (Map by Gary Asher and Les Joslin.)

Six

Pacific Northwest Region

Ranger Harold E. Smith (left) of the Pine Mountain Ranger District in the old Paulina National Forest, which was incorporated into the Deschutes National Forest in 1915, and other early Forest Service personnel such as forest assistant Will Sproat (right), often lived and worked under arduous conditions until they were able to build and occupy their first ranger stations. (Courtesy of Stephen G. Petit.)

Built by Forest Service ranger Emery J. Finch in 1907, thirty years before this photograph was taken, the Interrorem Ranger Station on the Duckabush River was the first ranger station in the Olympic National Forest and is one of the older Forest Service stations remaining in the Pacific Northwest Region. (Courtesy of the US Forest Service.)

Beginning in 1986, what by then was called the Interrorem Cabin was used by Forest Service volunteers. In 1994, the well maintained, square, one-story, three-room peeled-log cabin became an Olympic National Forest recreation rental cabin, which afforded visitors experiences approximating early forest ranger living. (Photograph by the author.)

In August 1911, Forest Service ranger Harvey Lickel raises the American flag at Gotchen Creek Ranger Station, built in the old Columbia National Forest in 1909. From 1909 to 1917, the cabin served as headquarters of the Mount Adams Ranger District. Columbia National Forest was renamed Gifford Pinchot National Forest in 1948 to honor the agency's founding forester. (Courtesy of the US Forest Service.)

Gotchen Creek Ranger Station, built along the primary "sheep driveway" by which tens of thousands of sheep annually entered the national forest bound for grazing allotments on the southern slopes of Mount Adams, was used for 84 years and is the oldest Forest Service structure in Gifford Pinchot National Forest. (Photograph by Rick McClure, courtesy of the US Forest Service.)

E.W. "Cy" Donnelly, first ranger of the Snow Mountain Ranger District, Ochoco National Forest—part of the Emigrant Creek Ranger District, Malheur National Forest, since 2008—built the first Allison Ranger Station cabin in 1911. He administered about 200,000 acres of national forest land from this one-room, pine log cabin—also known as the Donnelly Cabin—in use until 1925. (Photograph by the author.)

Today's historic Allison Ranger Station, still used as a guard station, is a compound of seven buildings built by the CCC in 1935 in the Pacific Northwest Region's distinctive Cascadian Rustic style. All these structures were beautifully restored in 2010 with American Restoration and Recovery Act of 2009 funding, and this warehouse was rehabilitated and returned to service as a modern station for a wildland fire engine crew. (Courtesy of the US Forest Service.)

The original Star Ranger Station was built in 1911 along the Applegate River in the old Applegate Ranger District of the Rogue River National Forest, which is now part of the Siskiyou Mountains Ranger District of the Rogue River–Siskiyou National Forest. Restored inside and out and furnished with period government furniture, it has seen recent use as a conference facility. (Photograph by the author.)

Star Ranger Station was home to legendary district ranger Lee C. Port from World War I to World War II. Across the road from the 1911 Star Ranger Station structure and the adjacent modern Star Ranger Station office facility is the second Star Ranger Station office building, constructed by the CCC in the 1930s. (Photograph by the author.)

Ranger Daniel D. Olin served at the Fish Lake Ranger Station in the Willamette National Forest in 1942. Set aside in 1906 for administrative use, the site was occupied from 1868 to 1906 by a way station for Santiam Wagon Road travelers. A log cabin ranger station was built there by the old Cascade National Forest in 1908 but was crushed by snow four years later. By the summer of 1914, the Santiam National Forest, partly carved out of Cascade National Forest, had replaced the crushed cabin with a two-room cabin and a six-stall barn. As time went on, other buildings were added. By the mid-1920s, the station was a group of very attractive log cabins that housed the forest supervisor during the field season, the fire dispatcher and his office, the two or three firemen stationed there, and the packer and pack animals. During the 1930s, the CCC added a remount depot that housed saddle and pack stock for fire management and wilderness management operations in Willamette National Forest—established by combining the Santiam and Cascade national forests in 1933—until 2005. (Courtesy of the US Forest Service.)

The 1921 dispatcher's cabin at Fish Lake Ranger Station was the old Santiam National Forest summer field headquarters for the many years in which legendary forester Charles Chandler "C.C." Hall was forest supervisor. His winter headquarters was in Albany, Oregon. (Courtesy of the US Forest Service.)

The restored dispatcher's cabin—complete with a replica "Santiam National Forest Field Headquarters" sign—remains in service at the historic Fish Lake Ranger Station and Remount Depot restored by Willamette National Forest and the nonprofit organization Friends of Fish Lake formed in 2010 by Forest Service retirees and others to ensure the historic station's survival and tell its story. (Photograph by the author.)

The beautiful Suiattle River and surrounding forests and mountains of the Mt. Baker–Snoqualmie National Forest provide a scenic and serene setting for historic Suiattle Guard Station, built in 1913 in the old Washington National Forest by legendary forest ranger Tommy Thompson to house a forest guard. Reached by horseback until the CCC constructed the Suiattle River Road in the late 1930s, and long used for forest management and protection work, this charming and well-built little cabin became a recreation rental. Thompson, born in England in 1884 and brought to the United States in 1885 by his parents, spent his life serving in today's Mt. Baker–Snoqualmie National Forest. Starting in 1904 as a fire guard in the Washington Forest Reserve, he was "among the small group that established the United States Forest Service in 1905," according to a plaque presented to him in 1955 by chief of the Forest Service Richard E. McArdle. Thompson retired in 1943 as district ranger of the Skagit Ranger District of Mt. Baker National Forest. (Courtesy of the US Forest Service.)

Forty miles as the crow flies south-southeast of Bend, Oregon, where the ponderosa pine forest gives way to the sagebrush sea, remote Cabin Lake Ranger Station was headquarters for five Deschutes National Forest district rangers between 1921 and 1945. A well drilled in 1916 to improve grazing in the Fort Rock Ranger District made the site an attractive location for a ranger station. (Courtesy of the US Forest Service.)

CCC crews constructed seven new Cabin Lake Ranger Station buildings between 1934 and 1936. District ranger Henry Tonseth, who led the district from 1934 to 1969—a regional record for district ranger tenure in one district—served 10 years there before his headquarters was moved to Bend in 1945. The station continued to be used as a guard station into the early 21st century. (Courtesy of Don Franks.)

As recreational use increased in the Deschutes National Forest, the Forest Service built Elk Lake Guard Station in 1929. Among many others into the 1990s, forest guard Clifford Wynkoop (second from left) provided information and assistance to visitors he met at the station and in the field during the 1947 and 1948 summers. (Courtesy of Fran Lattin.)

Restored between 1998 and 2001 by Forest Service personnel and Passport in Time volunteers, Historic Elk Lake Guard Station has served since 2002 as a Deschutes National Forest information station and interpreted heritage site where visitors receive information and assistance as they see how forest guards lived and worked. (Photograph by the author.)

One of only three log office structures built at Pacific Northwest Region ranger stations during the Great Depression, the new Silver Creek Ranger Station office completed in the Snoqualmie National Forest in 1933 nestles in a Douglas fir forest along the Mather Memorial Parkway just north of the White River entrance to Mount Rainier National Park. (Photograph by Fred W. Cleator, courtesy of the US Forest Service.)

A Forest Service ranger provides information to a Snoqualmie National Forest visitor at Silver Creek Ranger Station in the late 1930s. The centerpiece of the Silver Creek Guard Station Historic Site in the Mt. Baker–Snoqualmie National Forest, the well-preserved structure remains in service as Silver Creek Visitor Information Center. (Courtesy of the US Forest Service.)

Although the Forest Service first occupied the Rand Ranger Station site in the Siskiyou National Forest in 1909, the structures that now comprise the former Galice Ranger District headquarters and now used as an interagency visitor center were remodeled and built during the 1930s by the CCC. Added to the Rand Ranger Station ranger's residence, protective assistant's residence, and office already on site were seven new buildings to complete the compound. After World War II, tourism in the Rogue River Canyon increased, the Galice Ranger District's workload increased, and in 1958, a new ranger's residence and bunkhouse were built. In 1963, the Siskiyou National Forest moved the district's headquarters to Grants Pass, and the Rand Ranger Station saw only occasional use. With the designation of the Rogue River as a Wild and Scenic River in 1970, the US Department of the Interior acquired Rand Ranger Station for a Bureau of Land Management headquarters. Named the Smullin Visitor Center in 1996 after a broadcasting pioneer, historic Rand Ranger Station is within the Rand National Historic District managed by the Bureau of Land Management. (Photograph by the author.)

Tiller Ranger Station has been headquarters of five Umpqua National Forest ranger districts since 1918 and remains headquarters of the Tiller Ranger District. Nine historic buildings of the station's 27 structures were built in the Pacific Northwest Region's characteristic Cascadian Rustic style by the CCC between 1935 and 1942. (Courtesy of the US Forest Service.)

The 1935 house that long served as the district ranger's residence has been restored and furnished as the History House, dedicated to district rangers who have served at Tiller Ranger Station since its establishment. As a result of decades of district lumping and splitting, there have been many of them. (Photograph by the author.)

This painting by former Forest Service district ranger Evan Jones depicts the log cabin built at Clackamas Lake Ranger Station by ranger Joe Graham in 1906. By the time Graham finished his 24-year tour as district ranger at Clackamas Lake Ranger Station in the Mount Hood National Forest in 1930, forest use and management required a larger administrative complex. (Photograph by the author.)

Establishment of the CCC in 1933 and availability of New Deal funds allowed rapid construction of one of the Pacific Northwest's more beautiful Depression-era ranger stations now included in the Clackamas Lake Historic Area. The former office became a small visitor center. (Photograph by the author.)

The ranger's residence at historic Clackamas Lake Ranger Station was built in 1933 from a plan chosen by district ranger Francis Williamson from a book of Weyerhaeuser Timber Company plans entitled *Your Future Home*. It later became a recreation rental. A group called Friends of Clackamas Lake Historic Ranger Station helped the Forest Service preserve and interpret this National Register of Historic Places site. (Photograph by the author.)

With one unfortunate exception, historic Clackamas Lake Ranger Station appears about as it did during its prime in the 1930s and early 1940s when district ranger Otis J. "O.J." Johnson served there. That exception resulted from a May 26, 2003, fire that destroyed the protection assistant's residence, also known as the "Honeymoon Cabin," lived in by protection assistant Alton Everest and his wife as newlyweds, which was located behind the office. (Photograph by the author.)

On the windswept coast of southwestern Oregon, where the ancient Klamath Mountains meet the mighty Pacific Ocean, Gold Beach Ranger Station has braved the elements and served the public since the CCC built it for the Forest Service between 1935 and 1937. Today, it remains the picturesque headquarters of the Gold Beach Ranger District of the Rogue River–Siskiyou National Forest. (Courtesy of the US Forest Service.)

The original Gold Beach Ranger Station office building was erected at the entrance to the compound of nine Cascadian Rustic style buildings aesthetically united and functionally separated on a sloping marine terrace facing the ocean. In addition to the office were three residences, a crew house, a warehouse, a machine shop, an equipment storage building, and a gas and oil house. (Photograph by the author.)

The ranger's residence, built on the uppermost terrace of the Gold Beach Ranger Station site, enjoys the most privacy. As harmonious a group of government buildings to be found anywhere, all the original structures had clapboard siding, board and batten gables, and wood-shingled roofs. Accent masonry of distinctive beige rock, squared and coursed, is featured in foundations, chimneys, walks, and porches. Decorative ironwork, including pinecone door knockers and tree-shaped hinges, add to the effect. The landscape plan required many stone retaining walls to contain the terraces, as well as other built landscape features. Planted native trees, shrubs, and ground covers set off lawns that surround the office, the residences, and the crew house. This may seem excessive for a government project, but the price of the land and the original cost estimates for the nine buildings came to less than $20,000. That was not a high price to pay for a nine-structure ranger station compound now in its ninth decade of service—a project that made good use of the skills of otherwise unemployed and discouraged young men. (Photograph by the author.)

A rustic masterpiece in wood and stone, the Bly Ranger Station compound in remote Bly, Oregon, was built for the Fremont National Forest by the CCC and local men between 1936 and 1942. This group of nine administrative and residential buildings—augmented by a 1960s office structure to supplement the office building pictured—remains in service as a Fremont-Winema National Forest district ranger's headquarters. (Photograph by the author.)

All the Bly Ranger Station buildings—the office, the residences, this garage, the warehouse, and the gas and oil house—were built of native stone and timber in the Pacific Northwest Region's specified Cascadian Rustic style on a four-acre site in the town of Bly acquired for $625 in emergency relief funds. (Photograph by the author.)

The historic Monte Cristo Ranger Station—named for the mining community where gold was discovered in 1889—was headquarters of the Monte Cristo Ranger District of the Mount Baker National Forest from 1936 when its CCC builders completed the station's office and residence, until 1982, when the district was added to the Darrington Ranger District of the Mt. Baker–Snoqualmie National Forest. (Courtesy of the US Forest Service.)

Now called Verlot Public Service Center, the historic ranger station provides information services to about 50,000 Mt. Baker–Snoqualmie National Forest visitors annually. Additionally, the center houses interpretive exhibits that help connect visitors with the national forest, gateway communities, and the Mountain Loop Scenic Byway. (Courtesy of the US Forest Service.)

The small south-central Oregon town of Paisley is the home of the legendary 1.3-million-acre ZX Ranch, the Paisley High School Broncos, an annual mosquito festival, and the Depression-era Paisley Ranger Station built by locals and the CCC between 1937 and 1939 for the Fremont National Forest. The entire compound—an office building, a ranger's residence, a barn, a warehouse, a garage, and a gas house—cost the government less than $10,000. Although this "new" office building was added in 1963 and the historic 1930s buildings have been modified to accommodate current uses, the still-in-use district ranger's headquarters in the Fremont-Winema National Forest retains the feel and flavor of a pre–World War II ranger station. (Photograph by the author.)

Government Mineral Springs, 14 miles north of Carson, Washington, in the old Columbia National Forest—renamed Gifford Pinchot National Forest in 1949—has long been popular for the mineral springs many claim offer a plethora of medical benefits. In 1937, two years after the hotel that hosted visitors to these springs burned, the CCC built Government Mineral Springs Guard Station for Forest Service administration of the campground—significantly expanded for mineral springs visitors and other recreation activities—as well as for fire protection. In 1975, when giant Douglas firs and cedars posed a danger to campers, the campground was closed, and the guard station was abandoned. After that danger passed, Forest Service retiree Lloyd Musser urged restoration and reuse of the guard station. Gifford Pinchot National Forest heritage resource specialists planned the project and Mt. Adams Ranger District personnel supervised an AmeriCorps team, Pacific Northwest Forest Service Association retirees, Passport in Time volunteers, Skamania County Jail inmates, and contractors, who put the building right by 1972 when it became a popular recreation rental cabin. (Courtesy of the US Forest Service.)

The historic Glide Ranger Station office, built in the Umpqua National Forest by the CCC from the Wolf Creek CCC Camp in 1938, is a sturdy, one-story-with-basement, wood-frame structure that reflects the Pacific Northwest Region's signature Cascadian Rustic architectural style. The building served as the Little River Ranger District headquarters until 1964, when it was converted to a residence after a new office was built for the Glide Ranger District. That district, in 1984, was combined with the Steamboat Ranger District to form the current North Umpqua Ranger District, still headquartered in Glide but in a new office building. After renovation in 1990 restored the old office building to its original state, the Forest Service, in cooperation with the Roseburg Chamber of Commerce, in 1992 opened it as the Colliding Rivers Visitor Center in the "front yard" of the newer North Umpqua Ranger District headquarters to provide information to and interpretation of the area for visitors. (Courtesy of the US Forest Service.)

Adjectives such as "quaint" and "picturesque" are often applied to the Glacier Ranger Station built by the CCC and local men in the Mount Baker National Forest in 1938. On the east side of the settlement of Glacier, Washington, and not far south of the Canadian border, it remains a striking legacy of Great Depression public works programs in the national forests. (Courtesy of the US Forest Service.)

The historic native stone and timber building housed the old Glacier Ranger District headquarters for more than four decades. It remains in service as the Glacier Public Service Center operated jointly by the Forest Service and the National Park Service for Mt. Baker–Snoqualmie National Forest and North Cascades National Park visitors. (Photograph by the author.)

The value of Forest Service ranger stations for serving the public guided the 1930s decision to locate the former Chelan National Forest's new Chelan Ranger Station in the town of Chelan, Washington, at the southern end of Lake Chelan, rather than in a more isolated location within the national forest. District ranger Robert Foote welcomes visitors to his new station in July 1941. (Photograph by Fred W. Cleator, courtesy of the US Forest Service.)

The historic Chelan Ranger Station, now in the Okanogan-Wenatchee National Forest, remains in service as a visitor center. As year-round recreation has grown in the Lake Chelan area, the historic station has become a focal point of public activity and attention. (Photograph by the author.)

Early in the afternoon of June 15, 1942, school teacher John P. Robins—with wife Helen and their sons Dick, seven, and Dave, four—drove into the yard of the Paulina Lake Guard Station on the south shore of the larger of the two lakes within the Newberry Caldera in the Deschutes National Forest in Central Oregon. For the next 17 summers, Robins worked for district ranger Henry Tonseth as forest guard in the beautiful caldera that, in 1990, became the centerpiece of the Newberry National Volcanic Monument. Completed by the CCC during the spring of 1942, the guard station was the family's home during those summers. Restored to its original appearance, historic Paulina Lake Guard Station serves as a Newberry National Volcanic Monument summer visitor center. Established by Congress within the Deschutes National Forest, Newberry National Volcanic Monument is one of several national monuments administered by the Forest Service; most are administered by two US Department of the Interior agencies, the National Park Service and the Bureau of Land Management. (Courtesy of the US Forest Service.)

The original Leavenworth Ranger Station office, built in the 1920s, was the headquarters of the Wenatchee National Forest district ranger based in Leavenworth until it was replaced in 1939 by a new office building constructed by the CCC. The old office was converted into and used as crew quarters and later removed. (Courtesy of the US Forest Service.)

The current Leavenworth Ranger Station office in the Wenatchee National Forest is one of nine buildings of a ranger station compound built between 1937 and 1939. A beautiful example of the Cascadian Rustic architecture of Pacific Northwest Region structures of the Great Depression, it has been carefully maintained over eight decades and remains in service. (Courtesy of the US Forest Service.)

Seven

Alaska Region

Alaska, the largest state, harbors the largest national forests: the 16-million-acre Tongass National Forest and the 5-million-acre Chugach National Forest. Spanning southeastern Alaska, these temperate coastal forests are an extension of the Pacific Northwest rainforest. Over half of this acreage is on islands and on the coastal mainland where deep fjords and large glaciers hamper overland travel. And so, in Alaska, the Forest Service went to sea. If the Alaska Region, Region 10, has a distinctive ranger station architectural style, it is the naval architecture of its fabled ranger boats. (Map by Gary Asher and Les Joslin.)

Ranger boat *Chugach*, underway in the Wrangell Narrows, Tongass National Forest, Alaska, is the last of the wooden Forest Service ranger boats that, beginning in 1908, plied the 12,000-mile coastline of the nation's two largest national forests, the Chugach and the Tongass of southeastern Alaska. In the days before adequate aircraft services, these boats provided access to Alaska's dense island and coastal forests. (Courtesy of the US Forest Service.)

Designed for the Forest Service in 1925 and built in Seattle, Washington, ranger boat *Chugach* completed a major repair in Port Townshend, Washington, in 1995, and returned to her Petersburg, Alaska, home port. Ten years later, *Chugach* visited Portland, Oregon, for the Forest Service's September 2005 Centennial Reunion. (Photograph by the author.)

A historic Forest Service ranger station was saved from destruction when the community of Petersburg, Alaska, rallied in the late 1980s to preserve an important part of its heritage and gained a first-rate visitor center in the process. Completed in 1937 with Emergency Relief Appropriation Act of 1935 funds, the building had housed the Petersburg Ranger District office until 1985. (Courtesy of the US Forest Service.).

Its site once coveted for a parking lot, citizens saved the historic Petersburg Ranger Station building from demolition. The Forest Service entered into a partnership with the chamber of commerce to assist in renovation of the building and to share in the operation of the visitor center opened there in 1992. (Courtesy of the US Forest Service.)

Epilogue

This historic Forest Service ranger station office was built in 1933 at the Bridgeport Ranger Station in an old Mono National Forest ranger district east of the Sierra Nevada, which was transferred to the Toiyabe National Forest in 1945. After summer 1962, during which the author worked from this office, it was replaced by a new building and moved to Reese River Ranger Station in central Nevada. Decades later, when no longer in use, the author and Robert G. "Bob" Boyd of the High Desert Museum near Bend, Oregon, arranged its move to that museum. Restored in 2008, it helps interpret the role the Forest Service and the National Forest System played in the settlement and development of the American West. (Photograph by the author.)

BIBLIOGRAPHY

Boerker, Richard H.D. *Our National Forests*. New York, NY: The MacMillan Company, 1918.

Cayton, David W. and Caroline Metzler. *James G. Cayton: Pioneer Forest Ranger*. Rifle, CO: Cayton Ranger Station Foundation, 2009.

Cooley, Earl. *Trimotor and Trail*. Missoula, MT: Mountain Press Publishing Co. 1984.

Everest, Alton F. *Tales of High Clackamas Country*. Maupin, OR: Friends of Clackamas Lake Historic Ranger Station. 1993.

Gildart, Robert C. *Montana's Early-Day Rangers*. Helena, MT: Montana Magazine Inc., 1985.

Grosvenor, John R. *A History of the Architecture of the USDA Forest Service*. Washington, DC: US Department of Agriculture, Forest Service, 1999.

Hartig, Louis F. *Lochsa: The Story of a Ranger District and its People in Clearwater National Forest*. Seattle, WA: Pacific Northwest National Parks and Forests Association, 1989.

Joslin, Les. *Uncle Sam's Cabins: A Visitor's Guide to Historic Forest Service Ranger Stations of the West* (Rev. Ed.). Bend, OR: Wilderness Associates, 2012.

———. Images of America: *Deschutes National Forest*. Charleston, SC: Arcadia Publishing, 2017.

Lewis, James G. *The Forest Service and the Greatest Good*. Durham, NC: Forest History Society, 2005.

McClure, Rick, and Cheryl Mack. *For the Greatest Good: Early History of Gifford Pinchot National Forest*. Seattle, WA: Northwest Interpretive Association, 2008.

Moore, Bud. *The Lochsa Story: Land Ethics in the Bitterroot Mountains*. Missoula, MT: Mountain Press Publishing Company, 1996.

Perry, Walter J., and Les Joslin. *Walt Perry: An Early-Day Forest Ranger in New Mexico and Oregon, a Memoir*. Bend, OR: Wilderness Associates, 1999.

Pinchot, Gifford. *Breaking New Ground*. New York, NY: Harcourt, Brace, and Co., 1947.

———. *The Use of the National Forest Reserves*. Washington, DC: US Department of Agriculture, Forest Service, 1905.

Rakestraw, Lawrence, and Mary Rakestraw. *History of the Willamette National Forest*. Eugene, OR: US Department of Agriculture, Forest Service, 1991.

Riis, John. *Ranger Trails*. Richmond, VA: The Dietz Press, 1937.

Robins, Dick, and Dave Robins. *Seventeen Summers at Paulina Lake Guard Station*. Les Joslin, ed. Bend, OR: Wilderness Associates, 2006.

Shoemaker, Len. *Saga of a Forest Ranger*. Boulder, CO: University of Colorado Press, 1958.

Steen, Harold K. *The U.S. Forest Service: A History*. Durham, NC: University of Washington Press, 2004.

Williams, Gerald W. *The U.S. Forest Service in the Pacific Northwest*. Corvallis, OR: Oregon State University Press, 2009.

Visit us at
arcadiapublishing.com

CPSIA information can be obtained
at www.ICGtesting.com
Printed in the USA
LVHW072127061022
730124LV00007B/183